UNIQUE ANIMALS OF THE MOUNTAINS AND PRAIRIES

By Tanya Lee Stone

BLACKBIRCH PRESS
An imprint of Thomson Gale, a part of The Thomson Corporation

Detroit • New York • San Francisco • San Diego • New Haven, Conn. • Waterville, Maine • London • Munich

© 2005 Thomson Gale, a part of The Thomson Corporation.

Thomson and Star Logo are trademarks and Gale and Blackbirch Press are registered trademarks used herein under license.

For more information, contact
Blackbirch Press
27500 Drake Rd.
Farmington Hills, MI 48331-3535
Or you can visit our Internet site at http://www.gale.com

ALL RIGHTS RESERVED.
No part of this work covered by the copyright hereon may be reproduced or used in any form or by any means—graphic, electronic, or mechanical, including photocopying, recording, taping, Web distribution, or information storage retrieval systems—without the written permission of the publisher.

Every effort has been made to trace the owners of copyrighted material.

Photo Credits: Cover photo: © W. Perry Conway; © Jonathan Blair/CORBIS, 12; © Gary W. Carter/CORBIS, 6; © W. Perry Conway/CORBIS, 13 (bottom), 17, 18, 22, 23; © D. Robert & Lorri Franz/CORBIS, 3, 20, 21; © Dewitt Jones/CORBIS, 5; © Steve Kaufman/CORBIS, 16; © Karl Lehmann/CORBIS, 15; © Joe McDonald/CORBIS, 11, 13 (top); Photos.com, 8, 9, 10 (both), 19; Carol Polich/Lonely Planet Images, 14; © Galen Rowell/CORBIS, 7 ; © Kennan Ward/CORBIS, 7 (inset); Steve Zmina, 4

LIBRARY OF CONGRESS CATALOGING-IN-PUBLICATION DATA

Stone, Tanya Lee.
  Unique animals of the mountains and prairies / by Tanya Lee Stone.
    p. cm. — (Regional wild America)
  Includes bibliographical references and index.
  ISBN 1-4103-0445-0 (hard cover : alk. paper)
  1. Mountain animals—United States—Juvenile literature. 2. Prairie animals—Great Plains—Juvenile literature. I. Title II. Series: Stone, Tanya Lee. Regional wild America.
  QL155.S76 2005
  591.75'3'0973—dc22

2004011507

Printed in the United States of America
10 9 8 7 6 5 4 3 2 1

Contents

Introduction 5
Bear Country! 6
Big Buffalo 8
Lone Lynx 10
Jumpin' Jackrabbits 12
Balancing Bovines 14
Huge Herbivores 16
Pudgy Prairie Dogs 18
Bringing Back Black-Footed Ferrets 20
Prancing Prairie Chickens 22
Glossary 24
For More Information 24
Index 24

Introduction

The Mountains and Prairies region is home to a big part of the Rocky Mountains and the Great Plains. There are many national forests and parks in this region. The first national park in the United States was Yellowstone National Park, which opened in 1872. Yellowstone is in Wyoming and has areas that stretch into Montana and Idaho.

Although a variety of wildlife lives in the Mountains and Prairies region, some animals are especially well-known there.

An elk herd runs through the Montana wilderness, home to many nature preserves and parks.

Bear Country!

Black bears are not only black. They are seen in many shades of brown, caramel, reddish brown, and black. Black bears live in many parts of North America, but they are a common sight in the mountains and prairies. Feeding the bears in Yellowstone National Park quickly became a favorite American activity when the park first opened. (It is no longer allowed.)

Black bears are omnivores. They eat both plants and animals. They like fish and small mammals such as beavers and even young deer. They also like insects, such as ants and bees. But most of their diet is made up of fruits, nuts, leaves, and roots. They love acorns, blueberries, raspberries, and apples. Black bears do a lot of sleeping and playing.

A black bear climbs a tree to find food. These bears feed mostly on plants but sometimes eat animals.

Grizzly bears are brown. In fact, they are also called brown bears. They can be told apart from black bears by the hump above their shoulders. Grizzlies also have longer claws. The grizzly bear is the state mammal of Montana. Two hundred years ago, nearly 50,000 grizzlies roamed the plains of the United States. Today, there are less than 1,500 of them (outside of Alaska). Iowa, Montana, and Wyoming all take part in a grizzly bear recovery program to help protect these animals.

A mother grizzly shows her long claws (inset). Once common, grizzly bears are now scarce outside of Alaska.

Buffalo once roamed over most of North America but now live in protected areas.

Big Buffalo

In the 1800s, more than 30 million American buffalo wandered across North America. The largest groups roamed the plains from the Rocky Mountains to the Mississippi River. Today, there are a few hundred thousand left. They are often thought of as a symbol of this region. The buffalo, or bison, is the state mammal of Wyoming and Kansas. And they are a common sight on protected land in Montana, Nebraska, and North Dakota.

A bison is a big, furry animal with a large head and a hump on its back. They have an excellent sense of smell and keen hearing. Bison shed their thick winter coats in the spring. Both males and females have horns. Males can weigh more than 2,000 pounds (908kg). Their shoulders stand 5 or 6 feet (1.5 to 1.8m) off the ground. Females are a bit smaller.

These big animals are able to move quickly. They can run up to 25 miles (40km) per hour. They are good swimmers, too. When they travel in groups, they form a long line. These social animals also feed in groups. Bison are herbivores. They mainly eat grasses, grazing off and on throughout the day.

Despite its large size, a bison can move fast and even swim when necessary.

The lynx likes to hunt alone. It has long tufts of hair at the end of each ear (right).

Lone Lynx

The lynx belongs to the cat family. It is related to animals such as bobcats and mountain lions, which also live in this region. Lynx live in much of Canada and Alaska. But in the lower United States, the lynx is a threatened animal. This means it is likely to become endangered. The largest numbers of lynx in this region are found in forested areas of Idaho, Montana, Wyoming, and Utah. They are also scattered throughout parts of Colorado.

Using its large paws to run over the snow, this lynx pounces on its favorite dinner, the snowshoe hare.

    This medium-sized cat has long, thick fur on its body. Its long legs and big, furry paws help it walk in the snow. Its short tail often has a black tip. A lynx's unique ears have long tufts of hair that extend from the tip of each one. These cats are not very social. Male and female adults generally keep to themselves. But a mother lynx stays with her kittens for about a year.

    Lynx are nocturnal animals. This means they are mainly active at night. They are good hunters, with excellent eyesight and hearing. The lynx's main prey (an animal that is hunted by another animal) is the snowshoe hare. Like many cats, a lynx often hunts by stalking its prey quietly. Then it pounces and snatches up its meal!

Jumpin' Jackrabbits

Despite its name, the black-tailed jackrabbit is actually a hare! Hares and rabbits are related, but they are not the same. Hares are born with fur. And they can see and hop right away. Rabbits cannot do either at birth, and they are born without fur. Black-tailed jackrabbits are a common sight in this region. They also live in other parts of North America.

The biggest hare in North America, the black-tailed jackrabbit relies on its long legs for speed.

The black-tailed jackrabbit is the largest hare in North America. These animals weigh 3 to 7 pounds (1.4 to 3.2kg) and are about 19 to 25 inches (48 to 64cm) in length. Females tend to be a bit bigger than males. These hares have long legs. They also have very long ears that stand about 4 to 5 inches (10 to 13cm) tall. Black-tailed jackrabbits have black markings on their rump and tail. Many also have black markings on their ears. These hares are not social animals. Most travel alone. Females do not stay long with their young to care for them.

Black-tailed jackrabbits like open areas with plenty of shrubs and short grasses. They eat a variety of grasses and plants. They get most of the water they need from the plants they eat. Large shrubs offer jackrabbits places to hide.

Their predators (animals that hunt other animals for food) include hawks, eagles, foxes, bobcats, and mountain lions.

A black-tailed jackrabbit listens for danger with its long ears (above) to avoid predators like this eagle

Balancing Bovines

Sheep and goats are closely related. They both belong to a group called bovines. Bison and antelopes also belong to this group. The bighorn sheep is Colorado's state mammal. More bighorn sheep live in Colorado than anywhere else. They also live in all the other states of this region. Bighorn sheep are named for their horns. Both males and females have strong, curved horns. Male horns are longer, thicker, and more curved than female horns. Adult males are called rams. Adult females are called ewes. When the mating season begins, rams butt heads and fight to win the attention of ewes.

Bighorn sheep eat grasses and shrubby plants. They prefer grassy slopes that are near rocky cliffs. These sheep are expert climbers. They use this ability to get away from predators such as bear, bobcat, and lynx. When danger is near, they escape by climbing onto the steep, rocky areas. Their excellent eyesight helps them gauge exactly where to jump. And their strong legs and nimble feet help them balance.

An excellent climber, the bighorn sheep escapes predators by climbing onto rocky cliffs.

The mountain goat is another great climber that is a common sight in high elevations of this region. Their strong legs and flexible padded hooves help them grip craggy spots on a hillside. Male and female mountain goats have horns. Their thick, white coats keep them warm. Both bighorn sheep and mountain goats seek lower elevations when the snow gets deep in winter. Like bighorn sheep, mountain goats graze the hillsides. They eat grasses, mosses, lichens, herbs, and other plants.

The mountain goat uses its padded, cushioned hooves to grip rocky hillsides.

Huge Herbivores

The pronghorn antelope is found only in North America. It is actually not a true antelope, and is the only member of its family. A true antelope's horns do not fall off and do not have branches. But a pronghorn's do. Both the male and female have horns. The female grows horns that are shorter than the male's.

These animals are herbivores. They live in the prairies where there is enough sagebrush and other small plants for them to eat. Pronghorn antelopes have keen senses of sight, smell, and hearing. They have strong legs and are fast runners. In fact, pronghorn antelopes are the fastest mammals in North America. They can reach speeds of up to 50 miles (80km) per hour!

The pronghorn antelope lives only in North America. Both males and females have horns.

Rocky Mountain elk live in several western states. Males are recognized by their large antlers.

Rocky Mountain elk belong to the deer family. This elk is Utah's state mammal. They live west of the Rocky Mountains in several states. The largest numbers of Rocky Mountain elk are in Yellowstone. They eat grass, bark, and twigs. These large animals weigh between 400 and 900 pounds (182 to 409kg). Females are a bit smaller than males. Only male elk have antlers. The antlers grow to more than 5 feet (1.5m) wide. These massive antlers can weigh up to 30 pounds (14kg).

Pudgy Prairie Dogs

Prairie dogs are rodents. They are related to squirrels, chipmunks, and marmots. Prairie dogs are common in this region. They are also found in parts of the West and Southwest. Prairie dogs are small and plump. They weigh between 2 and 3 pounds (0.9 to 1.4kg). A prairie dog is about 12 inches (30cm) long. Its tail adds another 3 or 4 inches (7.6 to 10cm). Prairie dogs have short hair and are a cream or yellowish color.

Prairie dog families (left) live in large underground burrows. A prairie dog always stands guard outside the burrow, ready to warn of danger (right).

Prairie dogs dig burrows underground. They are known for their often large and complex burrow systems. These are sometimes called prairie-dog towns. There is a huge prairie-dog town in Badlands National Park in South Dakota. Prairie-dog burrows have tunnels that lead to different chambers for food storage, nesting, and sleeping. There are also toilet chambers and listening chambers in which prairie dogs listen for danger.

These animals live in family groups. More than one family may share a town and the groups help each other. There is always a prairie dog guarding the burrow. Prairie dogs are herbivores. There are many animals that prey on them. These include eagles, hawks, badgers, coyotes, and ferrets. If the prairie dog guarding the burrow senses danger, it sends an alarm call to the others. When the danger has passed, the guard makes an all-clear sound.

Bringing Back Black-Footed Ferrets

Black-footed ferrets are members of the weasel family. A black-footed ferret's main source of food is the prairie dog. They not only eat prairie dogs but also take over their burrows to use as their own! These ferrets are nocturnal. Prairie dogs are diurnal. That means they are active during the day and sleep at night. Black-footed ferrets hunt for prairie dogs at night, surprising them in their burrows.

The black-footed ferret is well named. It has cream-colored fur on its body, but its legs and feet are black. It also has black markings on its face and the tip of its tail. Black-footed ferrets are long and thin. A ferret's body is about 18 inches (46cm) long. Its tail adds another 6 inches (15cm) to its overall length.

A black-footed ferret peeks out of a burrow.

Black-footed ferrets used to live all over the Great Plains and parts of the Rockies. Today, they are endangered. These animals are being raised in captivity as part of a recovery program. Many have been released in prairie-dog territory in Wyoming, Montana, South Dakota, Utah, Colorado, and Arizona. Their numbers are still low, but there is hope that they will grow in the future.

Black-footed ferrets raised in captivity are being released into the wild in an effort to grow their population.

Prancing Prairie Chickens

The prairie chicken is a wild bird that lives on the grasslands and prairies, and is about the size of a domestic chicken. In the Mountains and Prairies region, prairie chickens are found in Kansas, Nebraska, and Colorado. They also live in grassland areas of Oklahoma and Texas. But their habitat keeps getting smaller as grassland is turned into cropland. Prairie chickens are sand-colored with brown markings. Their coloring helps them blend into the background.

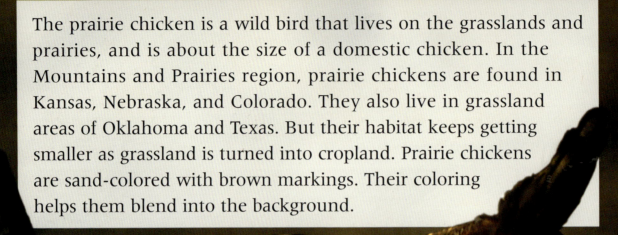

This prairie chicken puffs out his feathers and throat sacs to attract a female.

Prairie chickens have patches of bare yellow skin above each eye, and red throat sacs. Throat sacs are larger in males. When a male wants to attract the attention of a female, it puffs out its throat sac. Males also raise their neck and tail feathers, flap their wings, and jump up and down! After mating, the female lays 12 to 14 eggs in a grassy nest. When the eggs hatch in about a month, the chicks are fully formed. The mother stays nearby, but the chicks can feed themselves. Within a few weeks, they can fly.

These birds are omnivores. They mainly eat leaves, seeds, buds, and flowers. They also eat a lot of insects, such as grasshoppers. Hawks, eagles, owls, foxes, skunks, raccoons, and other animals prey on prairie chickens. Many of these animals will also take prairie chicken eggs from their nest.

There are many unique and wonderful animals that live in the Mountains and Prairies region. They all add to the richness and beauty of this area.

Two male prairie chickens fight over a female during mating season.

Glossary

Diurnal Asleep at night and active during the day.
Domestic Living with, or cared for by, people; tame.
Herbivore An animal that mainly eats plants.
Nocturnal Asleep during the day and active at night.
Omnivore An animal that eats plants and other animals.
Predator An animal that hunts another animal for food.
Prey An animal that is hunted by another animal.

For More Information

Barrett, Jalma. *Lynx* (Wildcats of North America). San Diego, CA: Blackbirch Press, 1998.

Jacobs, Liza. *Goats* (Wild Wild World). San Diego, CA: Blackbirch Press, 2003.

Patent, Dorothy Hinshaw. *Prairie Dogs*. New York: Clarion, 1999.

Swinburne, Stephen R. *Black Bear: North America's Bear*. Honesdale, PA: Boyds Mills Press, 2003.

Index

Antelope, 14, 16
Antlers, 17

Bears, 6–7, 14
Bighorn sheep, 14, 15
Birds, 22–23
Bison, 8–9, 14
Black-footed ferrets, 20–21
Bovines, 14–15
Brown bears, 7
Buffalo, 8–9
Burrows, 19, 20

Danger, 14, 19
Diurnal creatures, 20, 23

Ears, 11, 13
Elk, 17
Endangered animals, 10, 21

Ferrets, 19, 20–21
Fur, 11, 12, 20

Goats, 14
Great Plains, 5, 21

Hares, 12–13
Herbivores, 9, 16–17, 19
Horns, 9, 14, 15, 16

Jackrabbits, 12–13

Lynx, 10–11, 14

Mating, 14, 23
Montana, 5, 7, 8, 10, 21
Mountain goat, 15

National parks, 5
Nocturnal animals, 11, 20

Omnivores, 6

Prairie chickens, 22–23
Prairie dogs, 18–19, 20
Prairies, 16, 22
Predators/prey, 11, 13, 14, 19, 23
Pronghorn antelope, 16

Rocky Mountain elk, 17
Rocky Mountains, 5, 8, 17, 21
Rodents, 18–19
Running, 9, 16

Sheep, 14
State animals, 7, 8, 14, 17

Threatened animals, 10
Throat sacs, 23

Wyoming, 5, 7, 8, 10, 21

Yellowstone National Park, 5, 17